BUILDING BLOCKS OF CHEMISTRY

MIXTURES AND SOLUTIONS

Written by Cassie Meyer

Illustrated by Maxine Lee-Mackie

a Scott Fetzer company
Chicago

World Book, Inc.
180 North LaSalle Street
Suite 900
Chicago, Illinois 60601
USA

For information about other World Book publications, visit our website at **www.worldbook.com** or call **1-800-WORLDBK (967-5325)**.

For information about sales to schools and libraries, call 1-800-975-3250 (United States), or 1-800-837-5365 (Canada).

© 2023 World Book, Inc. All rights reserved. This volume may not be reproduced in whole or in part in any form without prior written permission from the publisher.

WORLD BOOK and the GLOBE DEVICE are registered trademarks or trademarks of World Book, Inc.

Library of Congress Cataloging-in-Publication Data for this volume has been applied for.

Building Blocks of Chemistry
ISBN: 978-0-7166-4371-5 (set, hc.)

Mixtures and Solutions
ISBN: 978-0-7166-4381-4 (hc.)

Also available as:
ISBN: 978-0-7166-4391-3 (e-book)

Printed in India by Thomson Press (India) Limited, Uttar Pradesh, India
1st printing June 2022

WORLD BOOK STAFF

Executive Committee
President: Geoff Broderick
Vice President, Editorial: Tom Evans
Vice President, Finance: Donald D. Keller
Vice President, Marketing: Jean Lin
Vice President, International: Eddy Kisman
Vice President, Technology: Jason Dole
Director, Human Resources: Bev Ecker

Editorial
Manager, New Content: Jeff De La Rosa
Associate Manager, New Product: Nicholas Kilzer
Sr. Content Creator: William D. Adams
Proofreader: Nathalie Strassheim

Graphics and Design
Sr. Visual Communications Designer: Melanie Bender
Sr. Web Designer/Digital Media Developer: Matt Carrington

Acknowledgments:
Writer: Cassie Meyer
Illustrator: Maxine Lee-Mackie/ The Bright Agency
Series Advisor: Marjorie Frank
Additional spot art by Samuel Hiti and Shutterstock

TABLE OF CONTENTS

Introduction 4
Properties of Mixtures 6
Heterogeneous Mixtures 8
Suspensions 10
Colloids 12
Homogeneous Mixtures..................... 14
Solutions 18
Solubility 20
Saturation..................................... 24
Separating Mixtures 28
Conclusion.................................... 34
Timeline 36
   Can You Believe It?!..................... 38
   Words to Know....................... 40

There is a glossary on page 40. Terms defined in the glossary are in type **that looks like this** on their first appearance.

INTRODUCTION

"Good afternoon! It's a perfect day in paradise and a great time for a meeting with Matter—that's me!"

"Matter is the name we give to all the different materials that make up the world. The warm sand, the blue ocean waves, and even the tropical breeze are all made of matter. You're made of matter, too!"

SMOOTHIE HUT

"Another name for a particular kind of matter is a substance."

PROPERTIES OF MIXTURES

Let's say that you throw a couple of substances together.

Did you make a mixture? Possibly!

A chemical reaction is a process by which one or more substances are converted into different substances.

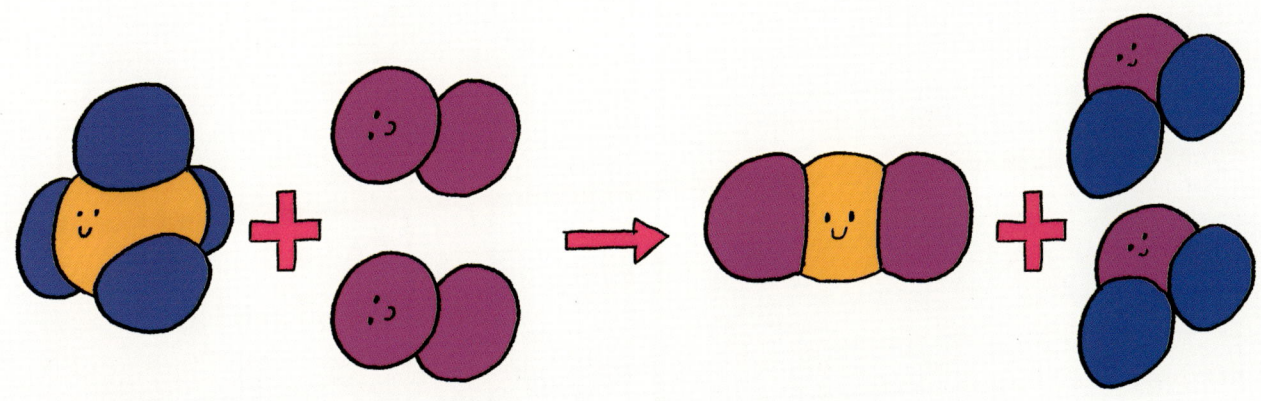

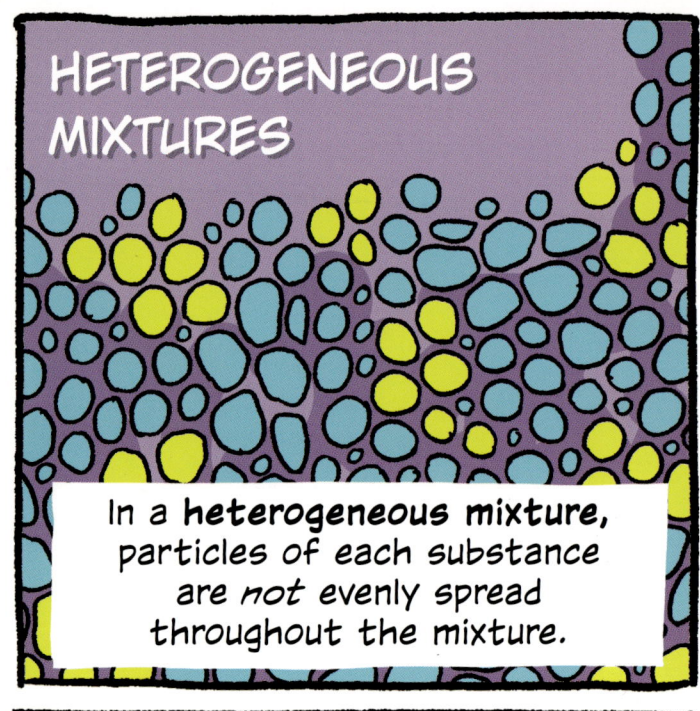

HETEROGENEOUS MIXTURES

In a **heterogeneous mixture**, particles of each substance are *not* evenly spread throughout the mixture.

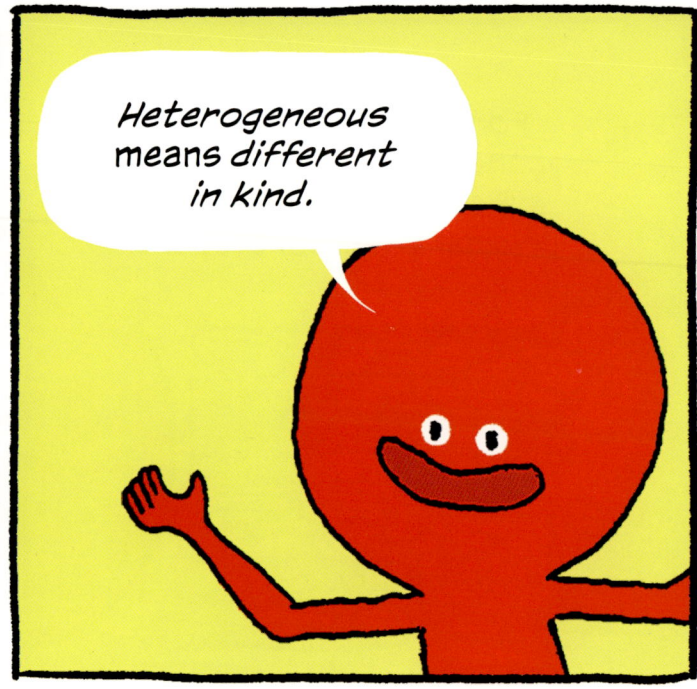

"Heterogeneous means *different in kind*."

Many everyday items around you are heterogeneous mixtures. They can be a mixture of solids such as this sandwich...

...or a mixture of liquids like a salad dressing...

...or a combination of solids and liquids like this pasta salad.

This ice and soda mixture contains all three states of matter! It's a mixture of gas bubbles, liquid pop, and solid ice.

Bon appétit!

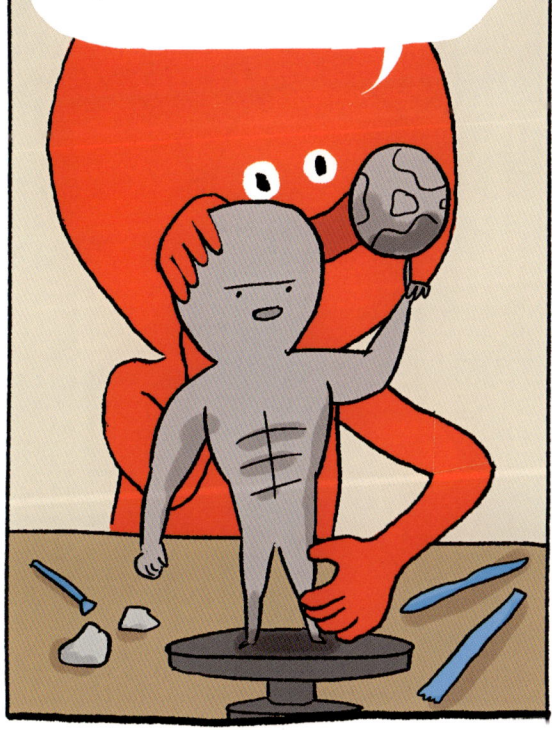

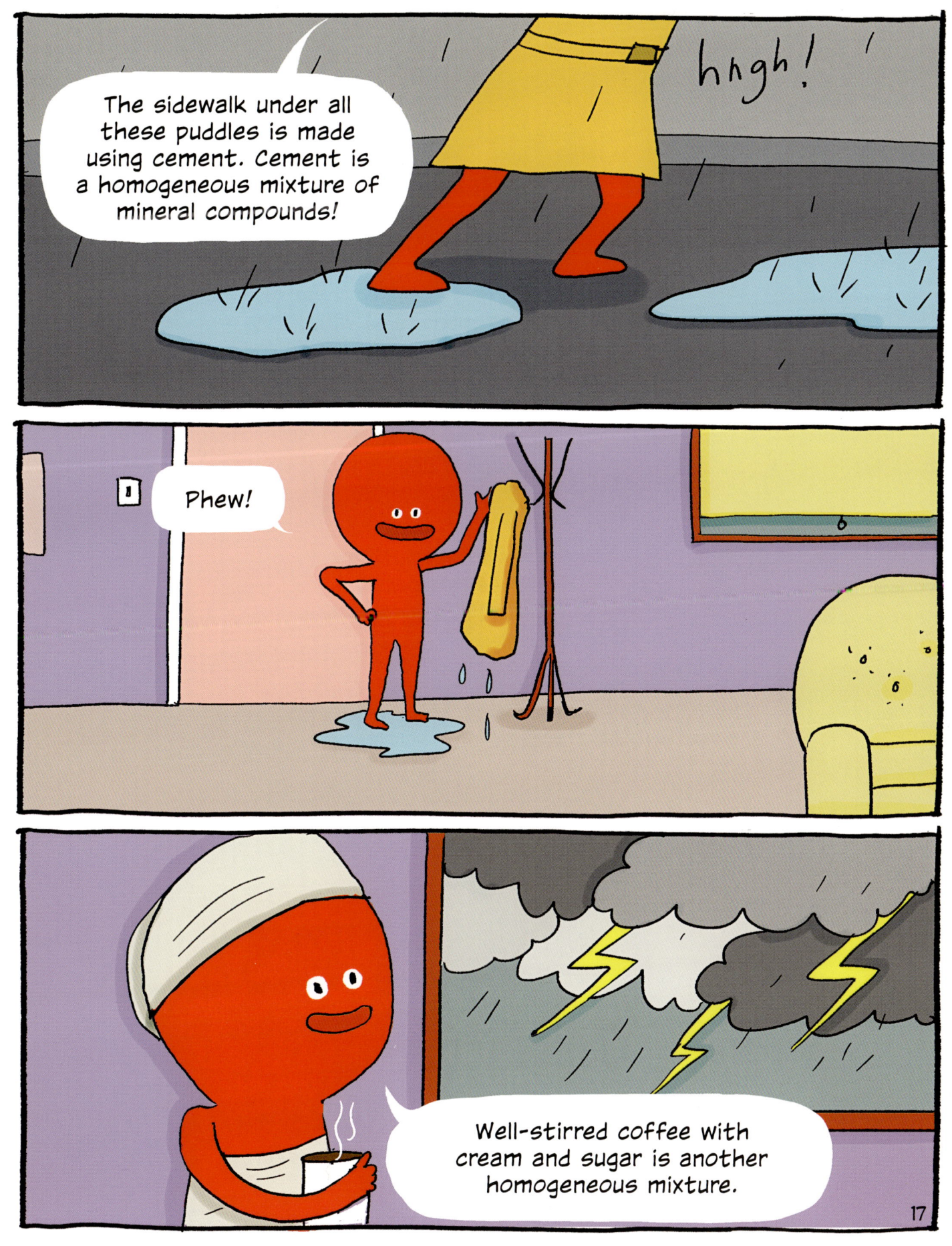

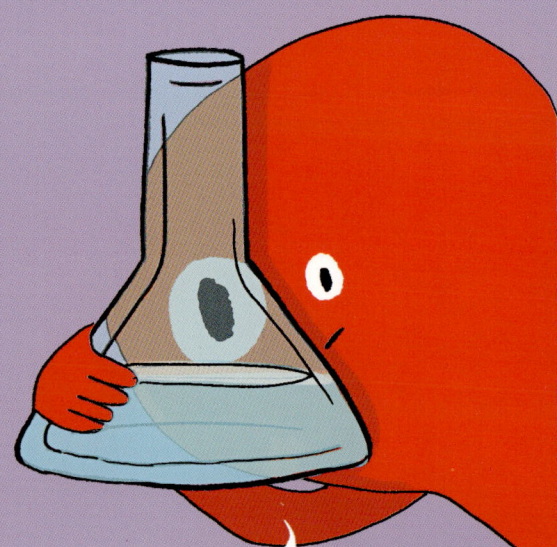

SOLUBILITY

"Sugar dissolves easily in the lemonade. This means that it is highly soluble."

Solubility is the ability of a solute to dissolve in a solution at a given temperature.

"Too sour? Guess it's back to the lab!"

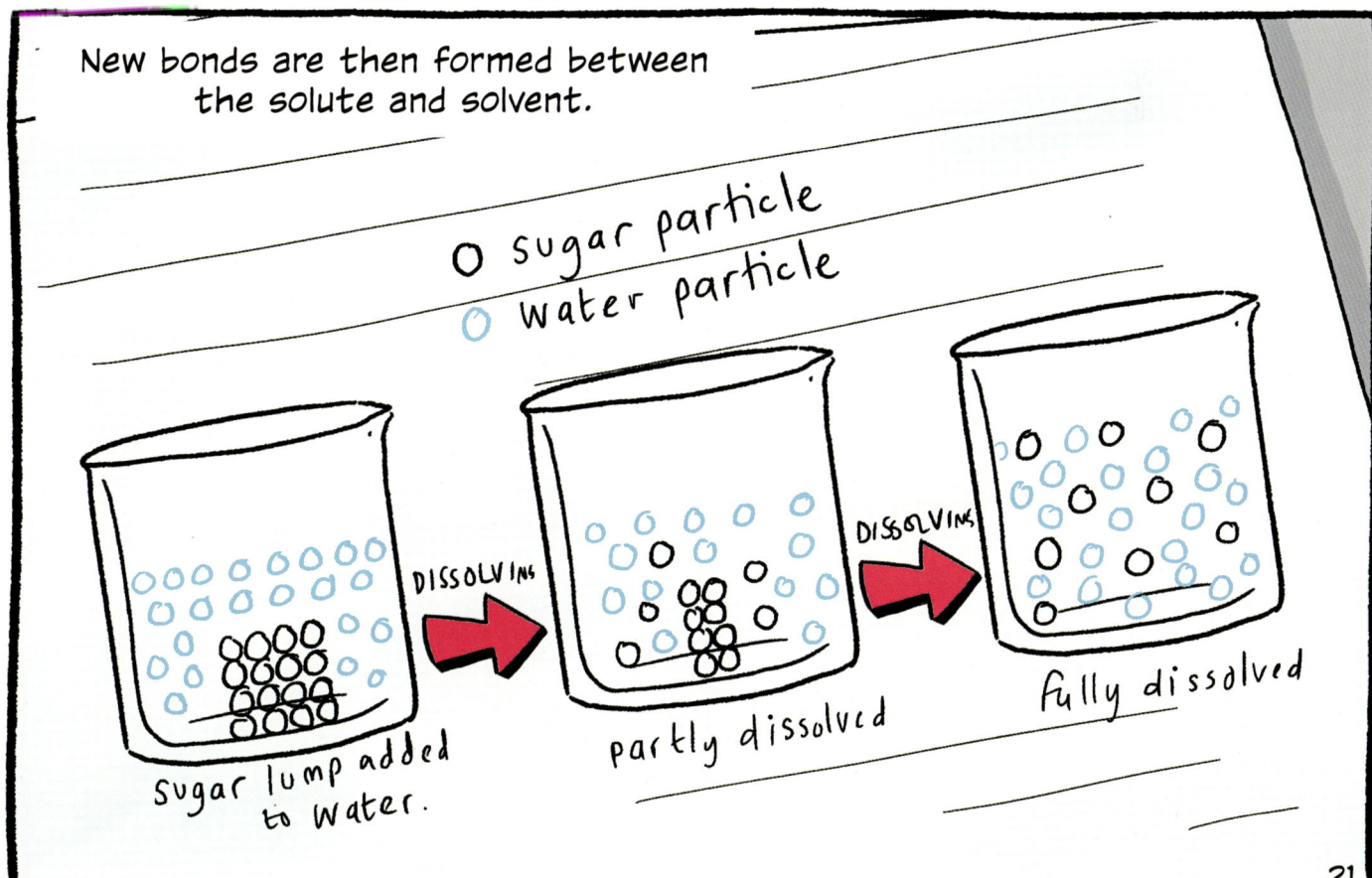

"For the solute to dissolve, the bonds that hold the solute together must be broken down."

New bonds are then formed between the solute and solvent.

○ sugar particle
○ water particle

sugar lump added to water. → DISSOLVING → partly dissolved → DISSOLVING → fully dissolved

The fat molecules tend to stick to themselves, so the water can't separate them.

Table salt has a solubility of about 35 grams per 100 milliliters of water at room temperature.

That means up to 35 grams of salt can dissolve in 100 milliliters of water.

Sugar has a much higher solubility than salt—about 200 grams per 100 milliliters of water.

23

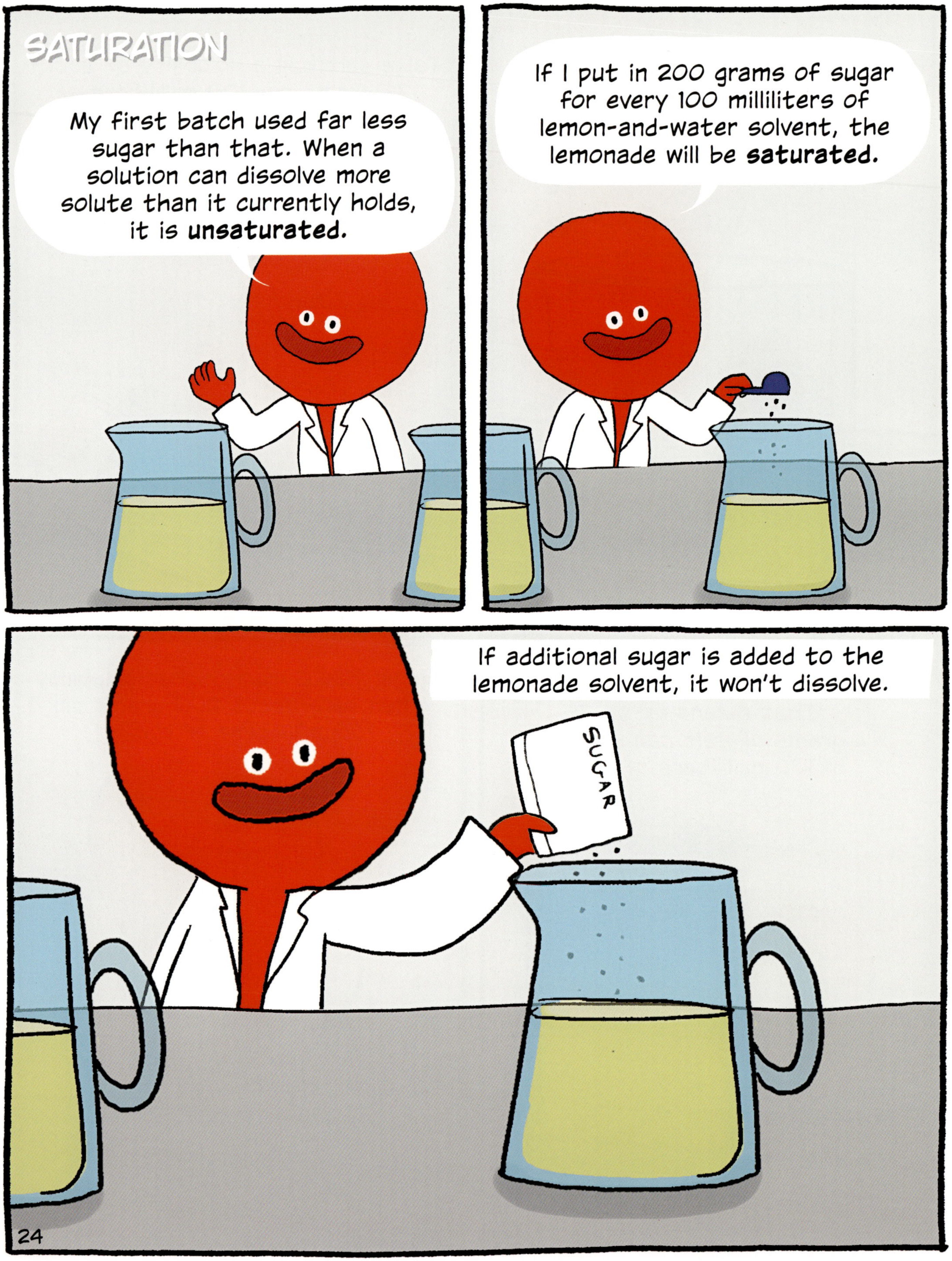

What happens to the solubility if we turn up the heat?

For most solid solutes, the higher the temperature, the more solute can be dissolved into the solvent.

When the solution is heated, the extra sugar dissolves.

The solution is now saturated at the higher temperature—but nobody wants to drink hot lemonade!

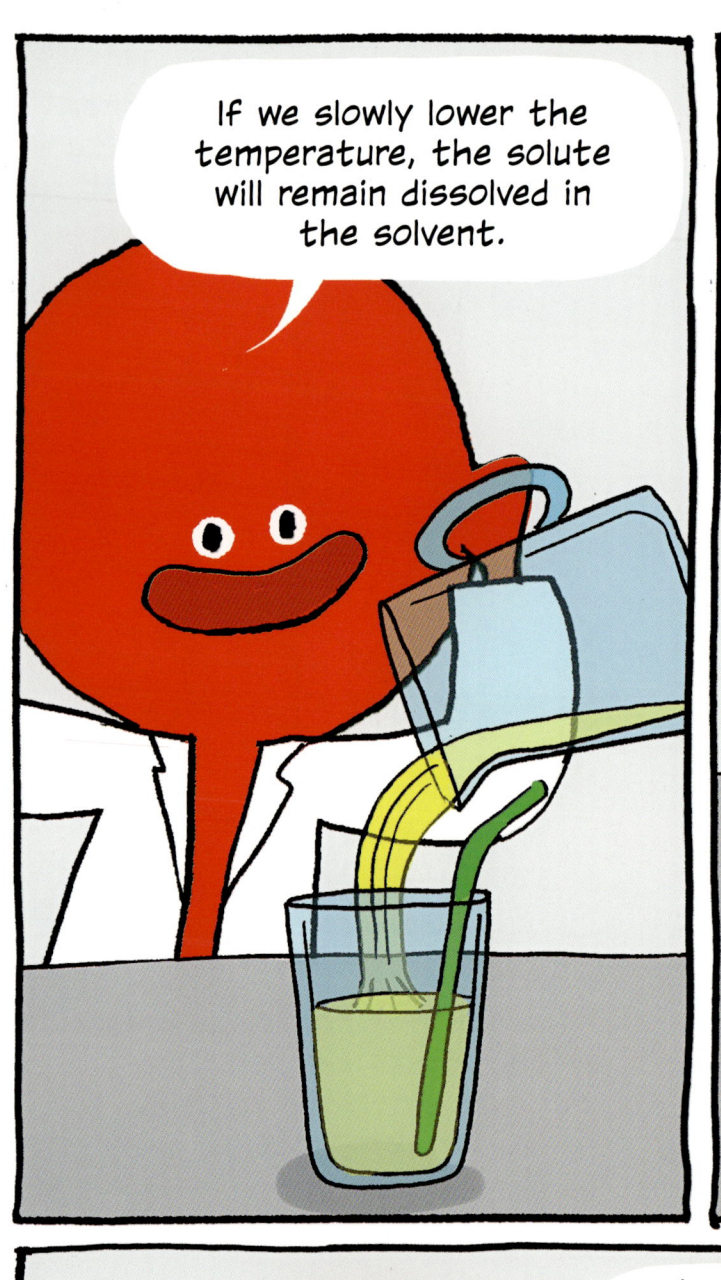

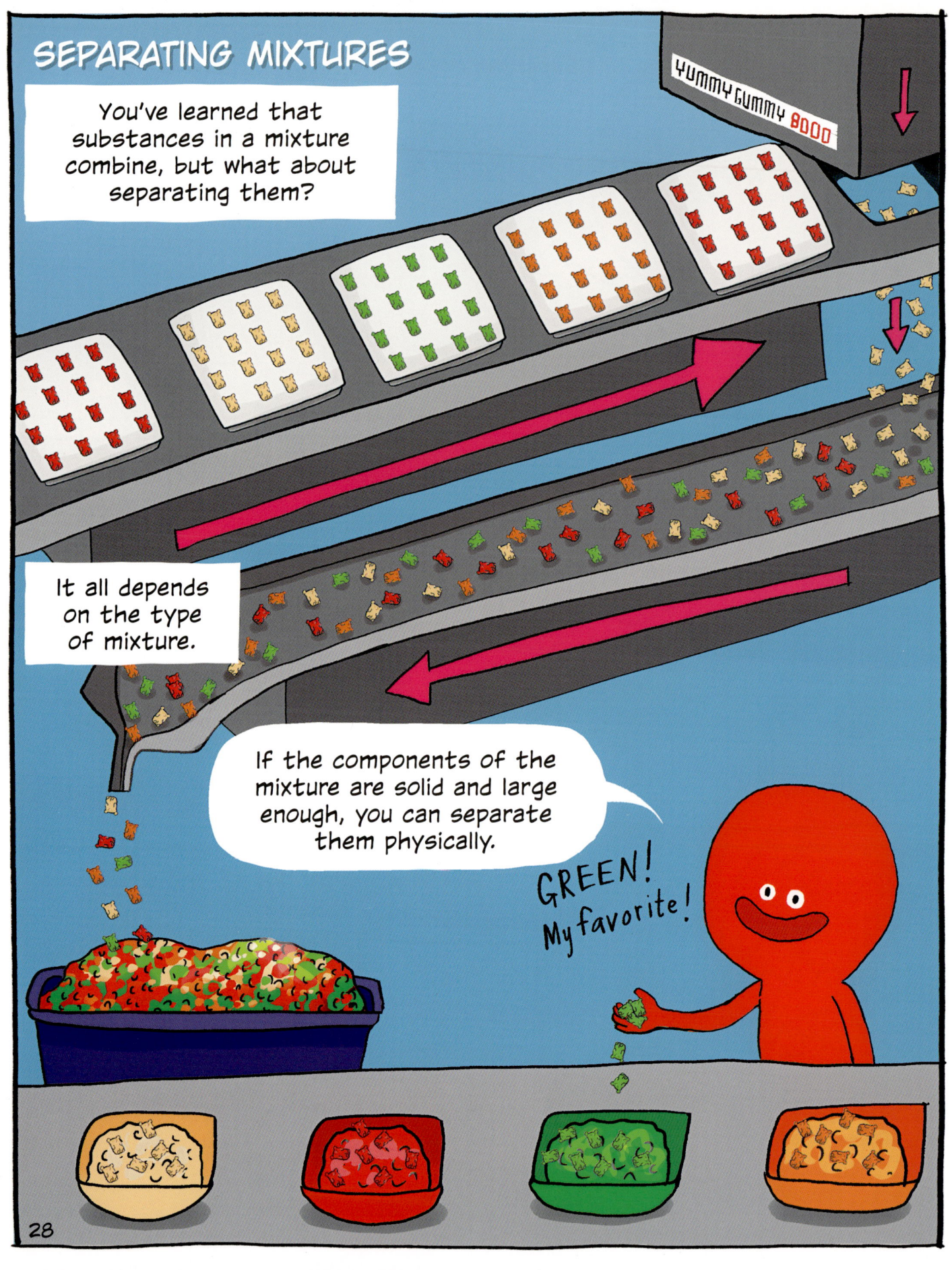

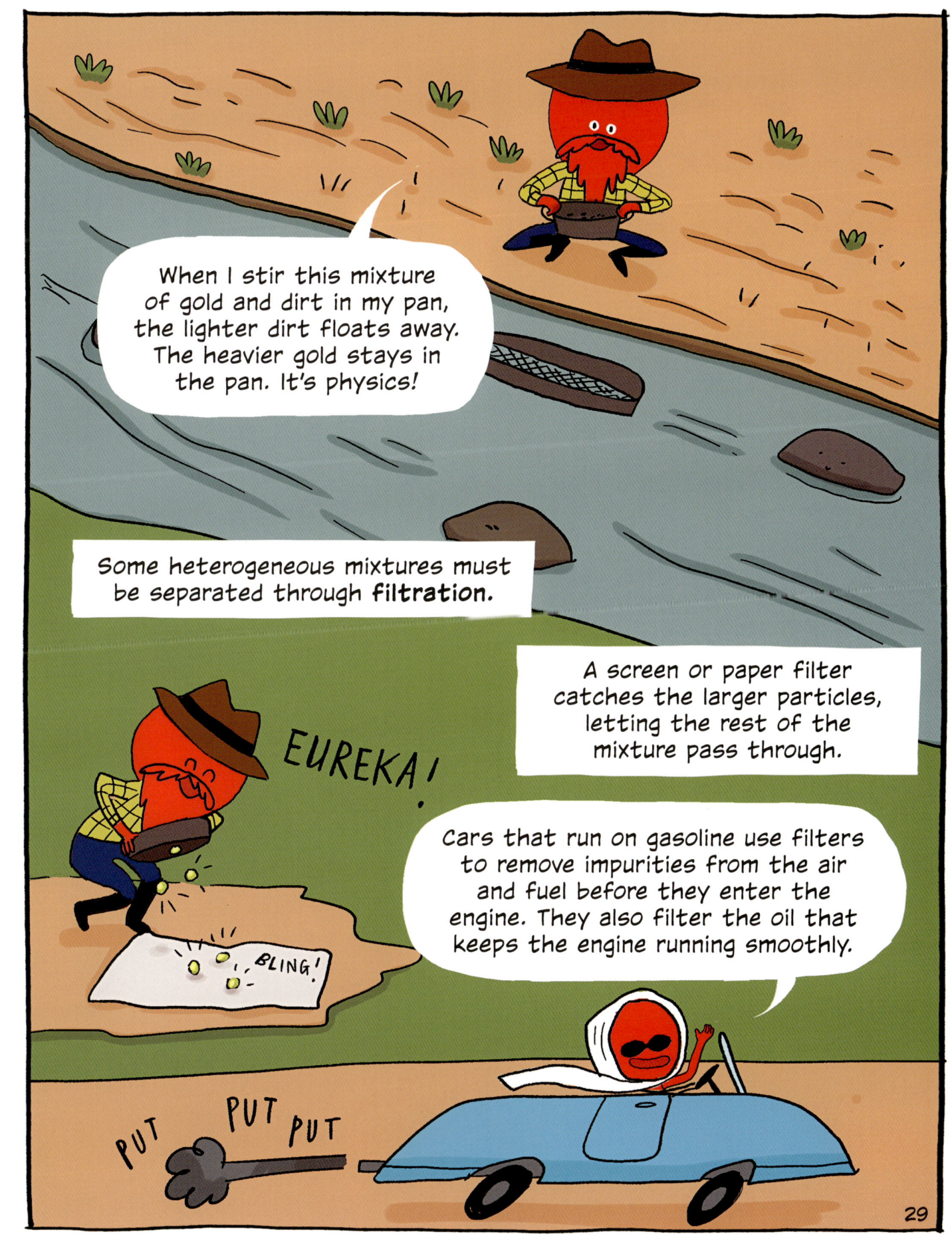

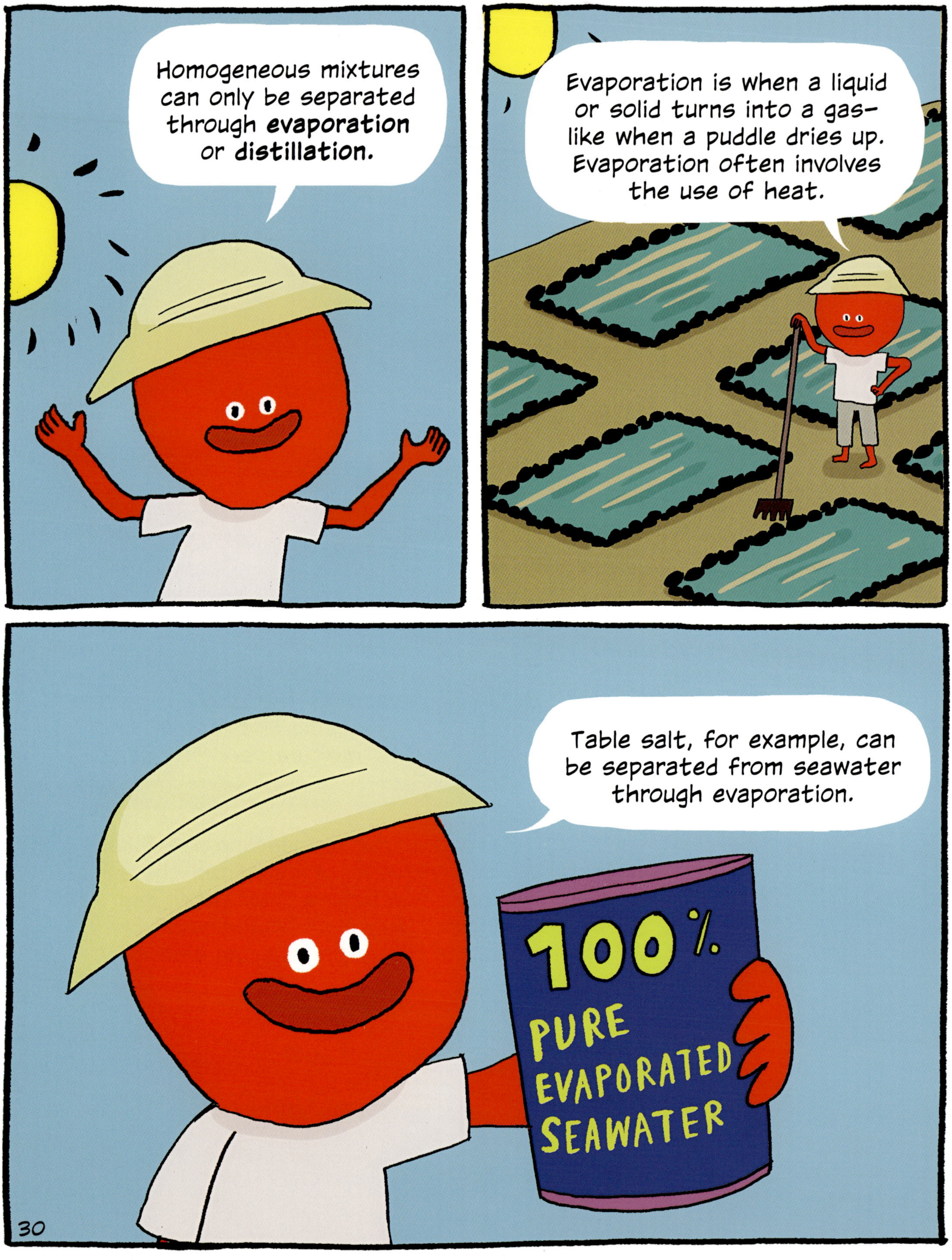

In places with hot climates, people build large, shallow beds to heat the salt water in the sun.

As the temperature of the salt water rises, the water evaporates into the air.

This leaves behind the salt portion of the mixture.

Simple evaporation is great if you want to harvest salt. But what if you need pure water?

Use distillation! Distillation separates substances from a solution through *vaporization* (turning into a gas).

A solar still is a simple way to purify water.

Heat from the sun causes the water to evaporate, leaving impurities behind.

A sheet of clear plastic traps the water vapor evaporating from the pool.

The water condenses on the plastic sheet.

A weight in the middle of the sheet causes the water to run toward the center.

Finally, the purified water drips down into the bucket.

32

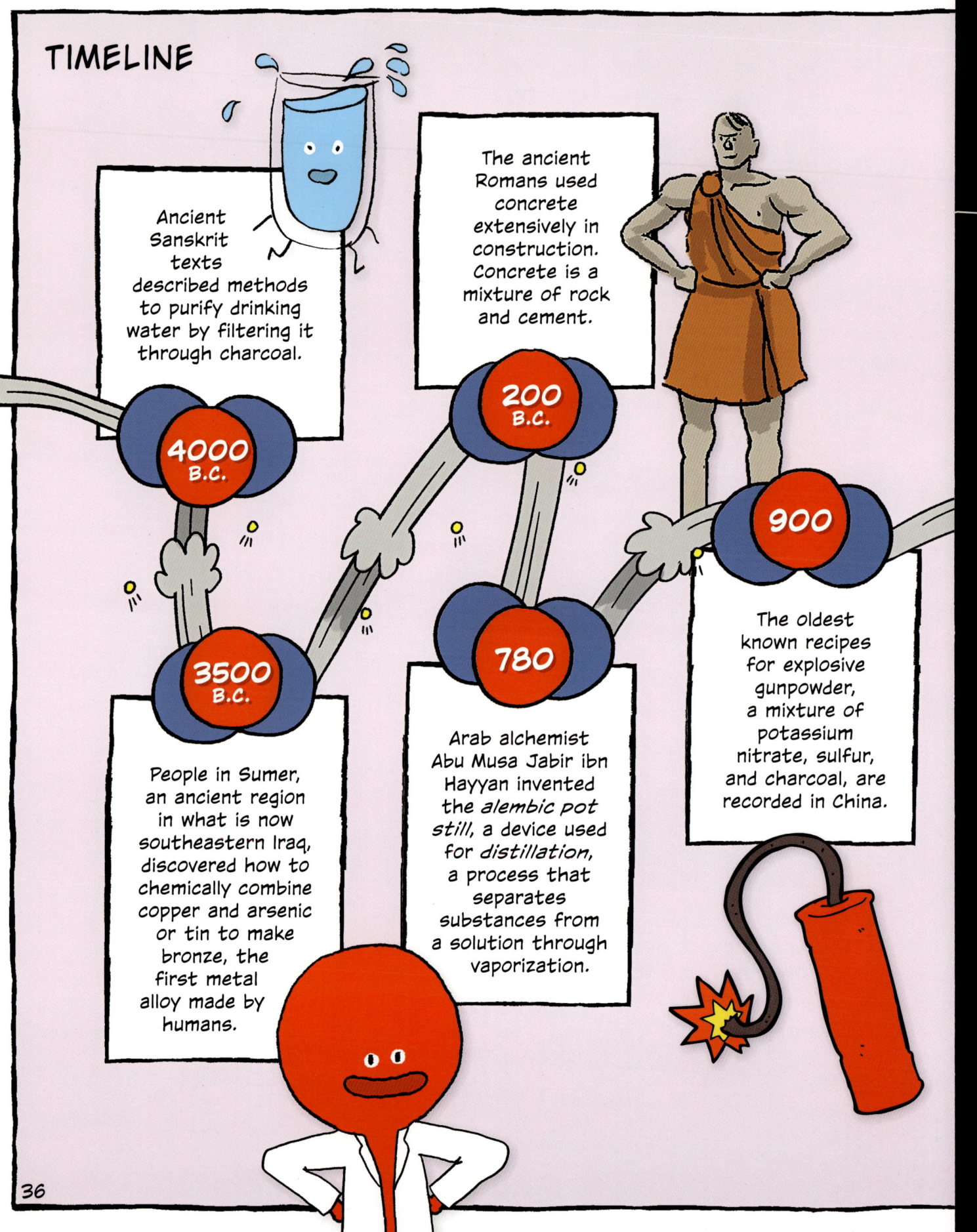

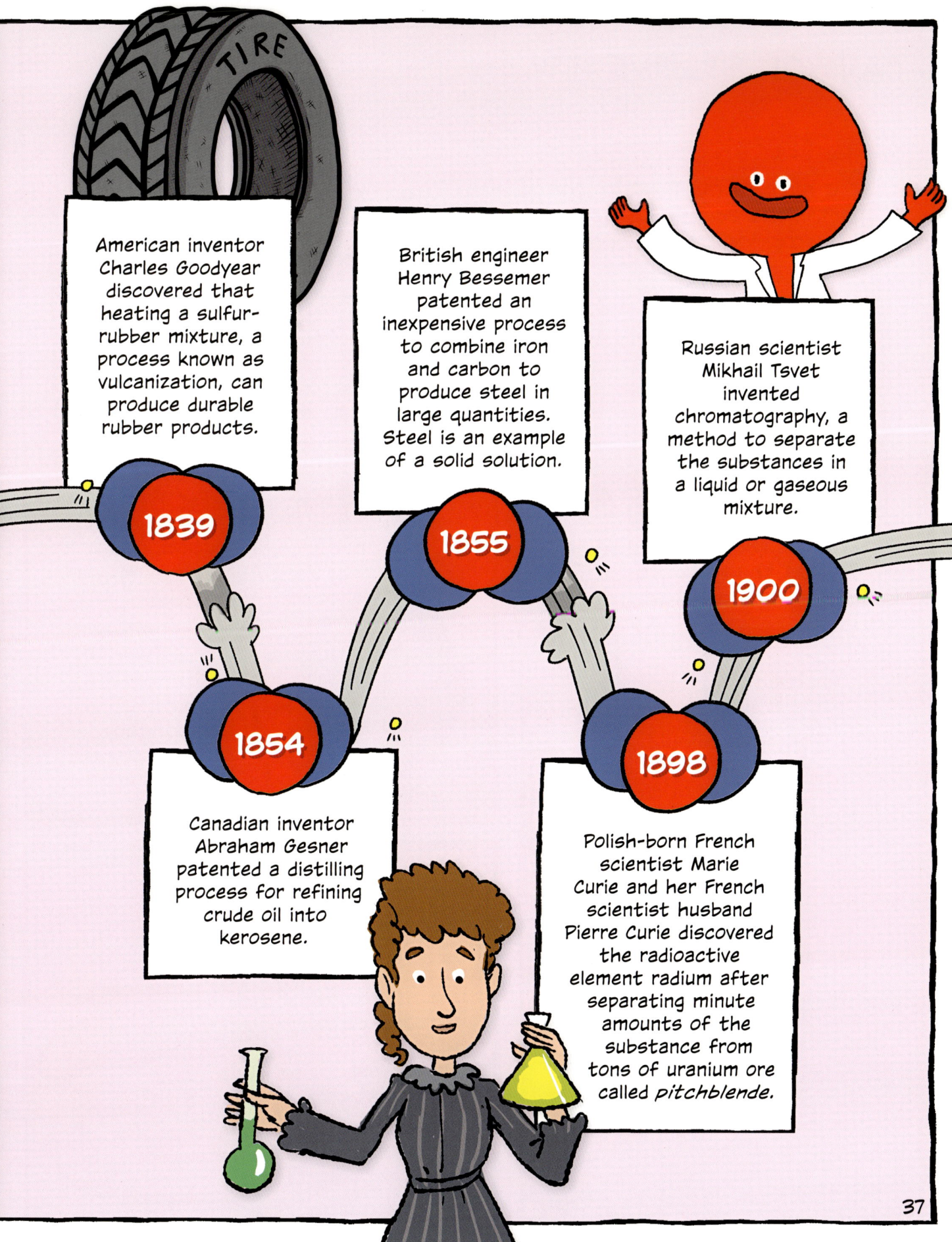

CAN YOU BELIEVE IT?!

Aqua regia, a mixture made up of one part nitric acid to three parts hydrochloric acid, is one of few substances **that can dissolve pure gold!**

Mix one liter of alcohol and one liter of water and the volume of the resulting liquid will be less than two liters! The molecules of alcohol and water **fit closer when mixed** compared to a pure sample of either liquid.

Only about 20 chemical elements occur in native (pure) form. These include carbon, copper, gold, helium, and nitrogen. Most chemical elements in nature are found combined with other elements.

A solution can be liquid, gas, or even a solid. **Stainless steel** is a solid solution of iron and carbon atoms with small amounts of oxygen.

The Pantheon in Rome, built in A.D. 126, is the **largest unreinforced concrete dome in the world.**
Concrete is a useful mixture of loose rock and cement.

Certain colloids suspended in liquid crystals can self-assemble into a variety of shapes and structures. Scientists are investigating these to make self-assembling *nanomachines* (extremely tiny machines).

The *Tyndall Effect* describes the effect of light scattering when shining through a colloid. This effect explains **why the sky and oceans appear blue.**
It is named for Irish physicist John Tyndall.

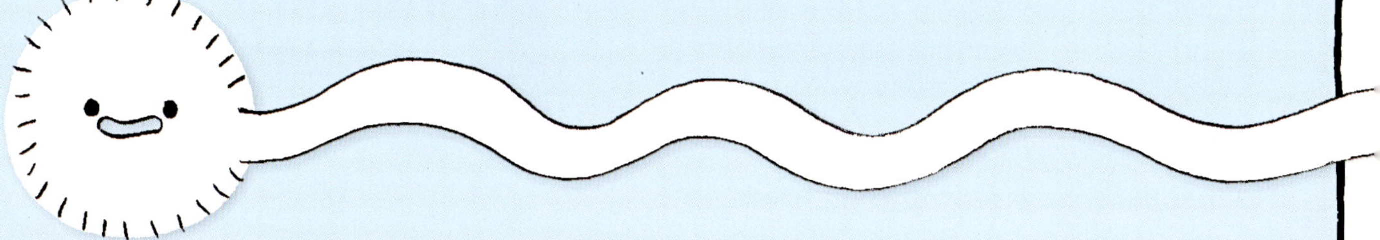

WORDS TO KNOW

chemical reaction a process by which one or more substances are chemically converted into one or more different substances.

colloid a material made of tiny particles of one substance that are distributed, but not dissolved, in another substance.

distillation a process that separates substances from a solution through vaporization.

evaporation the conversion of a liquid or solid to a gas.

filtration a process used to separate solids from liquids or gases using a filter that allows the fluid to pass through but not the solid material.

heterogeneous mixture a mixture in which the composition is not uniform throughout.

homogeneous mixture a mixture in which the composition is uniform throughout the mixture.

insoluble unable to be dissolved.

molecule two or more atoms chemically bonded together.

saturated describes a condition in which no more of a solute will dissolve in a solvent at a particular temperature and pressure.

solubility the ability of one substance to dissolve in another.

solute the substance that is dissolved by a solvent.

solvent a substance that dissolves another substance.

supersaturated describes a condition in which a solution contains more solute than it can normally hold when saturated.

suspension a mixture in which the particles of a substance separate from a liquid or gas slowly.

unsaturated describes a solution capable of absorbing or dissolving more of a substance.